All living things are connected to each other in many different ways. We call this the web of life.

Imagine you are walking into the woods. Here and there, you may see a deer, a rabbit, or a mouse. Look up at the tall trees with their flitting birds, moths, and butterflies. Glance down at the grasses, vines, and shrubs growing on the forest floor, many with berries or seeds on which forest animals feed. Spiders, worms, slugs, beetles, and hundreds of other creatures crawl around on the forest floor. The forest is filled with many kinds of plants and animals. Each is connected to the others in the life of the forest.

No plant or animal lives alone. Each depends in some way upon other plants and animals. And each has its own part in nature to play—producer, consumer, or decomposer.

Green plants use energy from the sun to make their own food from water, air, and soil. They are **producers**.

Animals eat the plants, or eat animals that eat the plants. They are **consumers**.

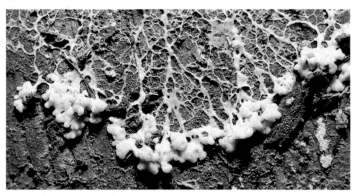

Other living things turn dead plants and animals into soil. They are **decomposers**.

All together, producers, consumers, and decomposers make up the web of life.

A tree is the largest of all plants. It is also the biggest producer in the web of life. Every spring, the tree grows new leaves, branches, flowers, and bark. At the same time, the trunk grows wider and stronger.

The leaves capture the energy of sunlight. They use the sun's energy to make food for the tree. The process is called **photosynthesis**.

Everything that grows on a tree provides food and energy for other living beings.

White-footed mice eat seeds, leaves, and fruit from the tree. These mice usually hide during the day and search for food at night.

Beetles often eat the leaves or bark of trees. This beetle is known as a "click beetle" because it makes a loud clicking sound if disturbed.

Anoles (uh-NOH-lees) are a kind of lizard.
They eat the insects they find on trees.

Animals of all kinds also find shelter and a place to hide in trees.

Robins build nests for their eggs in the branches.

Owls raise their young in a tree hollow.

The gray tree frog looks like a bump on the tree trunk. Can you guess how this helps it survive?

In summer, the flowers of a tree grow into fruits. The fruits of the oak tree are called acorns. Inside the acorns are seeds for making new oak trees.

Acorns are small. Yet many living creatures find food to eat and protection from enemies inside an acorn.

An acorn weevil drills a hole and then lays an egg inside an acorn growing on an oak tree.

In a week or two, the egg hatches into a larva (young insect). The larva eats the nutmeat inside the acorn and grows bigger and bigger.

When the acorn drops to the ground, the larva crawls out. It digs a hole in the soil and stays there until it becomes an adult weevil.

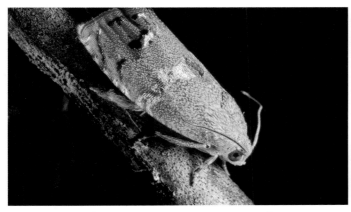

Meanwhile, a moth comes along. It lays an egg in the hole made by the weevil larva.

A caterpillar hatches from this egg and eats more of the acorn.

Tiny snails sneak into the acorn. They eat little bits of leftover nutmeat.

Ants lay eggs and care for the larvae inside the acorn.

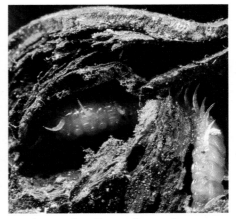

A centipede curls up in what's left of the nutshell.

Many wild birds and mammals eat acorns too.

Red squirrels can break the hard shells
easily with their chisel-like teeth.

The chipmunk puts acorns inside its
cheek pouches and scurries away.

Deer find a good supply of acorns in the woods where they live. Bears, foxes,
and wild turkeys are some other large animals that like to eat acorns.

Insects, disease, and severe weather kill many trees. But even a dead tree provides food and shelter for lots of animals.

Wasps lay their eggs in the bark.

A raccoon hides in a hole in the dead tree.

A flicker feeds four-day-old chicks in another part of the tree.

In time, every dead tree topples to the ground. Different kinds of plants start to grow on the fallen log.

Mosses are small green plants that grow on the log. They form a thick, soft carpet on the wood.

Mosses grow from tiny seedlike grains called spores. This moss spore case is releasing thousands of spores, which will be carried away by the wind.

Sometimes mushrooms appear overnight on a fallen tree. Mushrooms hold the spores of special kinds of plants called fungi (FUNJ-eye). Just one of these plants is called a fungus (FUNG-us).

The scarlet waxy cap, another kind of fungus, adds a bright dab of color to the log.

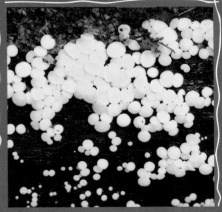

Yellow fairy cups glow against the dark background.

Insects and other animals come to the fallen tree to live and find food.

Termites eat their way through the rotting wood.

The short-tailed shrew looks for insects to eat.

The smooth green snake lives under the loose bark. But it comes out to hunt for food.

The box turtle creeps along the fallen tree.

Slimy slugs search for a place to lay their round eggs.

As time goes on, more of the tree decays. Decay is an important process in the web of life. Dead plants and animals break down into simple chemicals. The chemicals provide many of the nutrients that plants need to live and grow.

Most of the decay process is carried out by decomposers. These include bacteria, fungi, slime molds, and worms. The decomposers digest dead plants and animals and then return the digested remains to the soil.

Bacteria cause much of the decomposition. They are much too small to be seen without a microscope.

Fungi are plants, but they do not have leaves or chlorophyll (KLOR-uh-fill), the substance that gives green plants their color and helps them make their own food. Instead, fungi get their food from their surroundings.

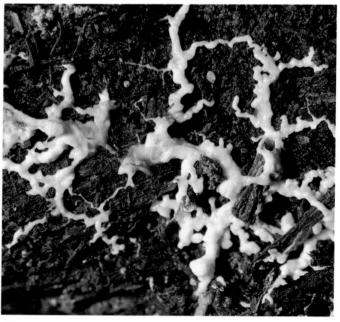

Slime molds are another kind of fungus. They slowly creep along as they feed and help to decompose the wood.

Earthworms feed on fallen logs. They also burrow into the ground, loosening the soil and helping new plants to grow.

The decomposers that feed on the log grow and multiply. The decay process speeds up. After several years, much of the wood has rotted away.

The wood becomes soft and spongy. Soon it looks, feels, and smells like soil.

After fifteen years, the fallen tree has become part of the forest floor. New plants start to grow here.

In the rich soil created from the decomposing tree, the seed from an acorn has started to grow. Eventually, the little plant will become a tall oak tree, reaching for the sun. The tree will grow leaves, flowers, fruit, and seeds. Many different animals will come to the tree for food and shelter. And the web of life will go on and on.